Harry Clary Jones

The Freezing-Point

Boiling-Point and Conductivity Methods

Harry Clary Jones

The Freezing-Point
Boiling-Point and Conductivity Methods

ISBN/EAN: 9783337275570

Printed in Europe, USA, Canada, Australia, Japan

Cover: Foto ©berggeist007 / pixelio.de

More available books at **www.hansebooks.com**

...THE...

Freezing-Point, Boiling-Point,

——AND——

Conductivity Methods

—BY—

HARRY C. JONES,

INSTRUCTOR IN PHYSICAL CHEMISTRY IN JOHNS

HOPKINS UNIVERSITY

EASTON, PA.:
CHEMICAL PUBLISHING CO.
1897

(All rights reserved.)

COPYRIGHT, 1897, BY EDWARD HART.

PREFACE

I have been impressed, in teaching the physical chemical methods in the laboratory, with the fact, that there is no readily accessible place in which they are treated satisfactorily from both the standpoint of theory and of practice. In the text-books, the theoretical side is developed, and usually without sufficient attention to the details of manipulation, to enable them to be applied successfully in the laboratory. In the laboratory manuals, on the other hand, these methods are often treated largely from the mechanical side, and their theoretical bearing might thus be lost sight of.

The physical chemical methods, which find most frequent application in the laboratory, are probably those based upon the lowering of the freezing-point, and the rise in the boiling-point of a solvent, produced by a dissolved substance, and the electrolytic conductivity of solutions of electrolytes. It is my chief object in preparing this little work to give an account of the operations involved in carrying out these methods in the laboratory. But since the mere mechanical application of any scientific method is a matter of comparatively little significance, I have aimed to give, also, enough of the theoretical ground on which each of them rests, to enable the student to work with them intelligently, and to see clearly their scientific significance and use.

HARRY C. JONES.

CONTENTS

PART I

THE FREEZING-POINT METHOD

	PAGE
Theoretical Discussion	1
Early History	1
Work of Raoult	1, 2
Molecular Lowering for Different Solvents	3
Molecular Lowering in Aqueous Solutions	4
Theory of Electrolytic Dissociation	5
Calculation of the Molecular Lowering	6, 7
Experimental Verification	8
Calculation of Molecular Weights from Lowering of Freez-Point	8, 9
The Application of the Freezing-Point Method to the Determination of Molecular Weights in Solution	9
The Apparatus of Beckmann	10, 11
Carrying out a Determination	11-13
Correction for the Separation of Ice	13, 14
The Application of the Freezing-Point Method to the Measurement of Electrolytic Dissociation	14
The Method of Calculating Dissociation from Lowering of Freezing-Point	15, 16
The Method of Work	16
The Apparatus of Jones	17, 20
Comparison of the Results with the Dissociation from Conductivity Measurements	21

PART II

THE BOILING-POINT METHOD

	PAGE
Theoretical Discussion	23
Historical	23, 24
Work of Raoult	24, 25
The Relative Lowering of the Vapor-Tension	26
Calculation of Molecular Weights from Lowering of the Vapor-Tension	27
Beckmann's Work on Rise in Boiling-Point	27, 28
Calculation of Molecular Weights from Rise in the Boiling-Point of Solvents	28
Values of the Constants for Solvents	29
Relations between Boiling-Point and Freezing-Point Methods	29, 30

The Application of the Boiling-Point Method to the Determination of Molecular Weights in Solution ... 30

The Apparatus of Beckmann	31-33
The Apparatus of Hite	33-35
The Apparatus of Jones	34-36
Carrying Out a Determination	36-39
Correction for Separation of Vapor	39
Results of Measurements	40, 41

PART III

THE CONDUCTIVITY METHOD

Two Classes of Conductors	42
Electrolytes and Non-Electrolytes	42
Specific Conductivity	43, 44
Molecular Conductivity	44
Dissociation Measured by Conductivity Method	45, 46
Determination of μ_∞	46-50

PAGE

The Application of the Conductivity Method to the Measurement of Electrolytic Dissociation 50
The Apparatus Employed 50–52
Calculation of the Molecular Conductivity............. 52, 53
Temperature Coefficient of Conductivity.................. 54
The Ostwald Thermoregulator............................ 55
Calibrating the Wire 56–58
Carrying Out a Conductivity Measurement 58
Determination of the Cell Constant...................... 59
Precautions 60
Correction for the Conductivity of Water.............. 60, 61
The Purification of Water 61–63
Substances to be Used.................................. 64

PART I

THE FREEZING-POINT METHOD

Theoretical

It has long been known that when a solid is dissolved in a liquid, the freezing-point of the solution is lower than that of the solvent. The first quantitative relation we owe to Blagden,[1] who pointed out that the lowering of the freezing-point of water, produced by different amounts of the same substance, was proportional to the amount of substance present. This same fact was rediscovered much later by Rüdorff.[2] A marked advance was made by Coppet,[3] who dealt with comparable, rather than with equal amounts of different substances. He used quantities of different substances which bore to one another the same relation as their molecular weights, and found that such quantities, of substances which are chemically allied, produce very nearly the same lowering of the freezing-point of any given solvent. In a word, the lowering of the freezing-point of a solvent by a dissolved substance, is proportional to the number of parts of the substance present.

This is about what was known when the problem was taken up by Raoult, and it is to him more than to any other that we owe the present development of the freezing-point method. He investigated water solutions of organic compounds, and found that the lowering produced by molecular quantities was very nearly a constant. He used other solvents, such as benzene, and

[1] Phil. Trans., 78, 277.
[2] Pogg. Annalen, 114, 63; 145, 599.
[3] Ann. chim. phys., [4], 23, 366.

found that comparable quantities of dissolved substances produced the same lowering of the freezing-point. His investigations included nitrobenzene, ethylene bromide, formic and acetic acids, and in each solvent a large number of substances were dissolved. He was thus in a position not only to compare the lowerings produced by different substances in the same solvent, but the lowerings in different solvents.

As the result of this work, Raoult attempted the following generalization.

One molecule of any complex substance dissolved in one hundred molecules of a liquid, lowers the freezing-point of the liquid by nearly a constant amount, which is 0.62°. This has been shown not to hold rigidly.

When a gram-molecular weight of any substance is dissolved in say 100 grams of a solvent, the lowering of the freezing-point of the solvent is a constant, regardless of the nature of the substance, provided that there is no aggregation of the molecules of the substance, and no dissociation. This was shown to hold approximately for a large number of substances, and for several solvents by Raoult.[1] The molecular lowering, which is the lowering produced by a gram-molecular weight of the substance in 100 grams of the solvent, was calculated by him thus:

If g grams of the substance are dissolved in 100 grams of the solvent, if m is the molecular weight of the substance, and Δ the lowering of the freezing-point of the solvent produced by the presence of g grams of the substance, then the molecular lowering is calculated from the formula:

$$\text{Molecular lowering} = \frac{m\Delta}{g}.$$

[1] Ann. chim. phys. [6], 2, 66.

THE FREEZING-POINT METHOD

A few results will show the values of the molecular lowering for different solvents.

Solvent, Acetic Acid.

	Molecular lowering.
Methyl iodide	38.8
Aldehyde	38.4
Acetone	38.1
Benzoic acid	43.0
Ethyl alcohol	36.4
Acetamide	36.1
Stannic chloride	41.3
Carbon disulphide	35.6
Sulphuric acid	18.6
Hydrochloric acid	17.2

Solvent, Benzene.

Methyl iodide	50.4
Anthracene	51.2
Ether	49.7
Acetone	49.3
Chloral	50.3
Stannic chloride	48.8
Methyl alcohol	25.3
Ethyl alcohol	28.2
Benzoic acid	25.4

Solvent, Water.

Methyl alcohol	17.3
Cane-sugar	18.5
Acetamide	17.8
Chloral hydrate	18.9
Milk-sugar	18.1
Acetone	17.1
Hydrochloric acid	39.1
Nitric acid	35.8
Sulphuric acid	38.2
Sodium hydroxide	36.2
Potassium chloride	33.6

A careful study of these results will bring out some interesting facts. The value of the molecular lowering of acetic acid and of benzene is very nearly a constant for each solvent. This is true for a large number of substances of the general type of most of those given above, *i. e.*, non-electrolytes. There are, however, exceptions for these solvents. In the case of acetic acid, there are a few substances known which, like sulphuric acid, give a molecular lowering of only one-half that produced by the non-electrolytes. In benzene there are also a few exceptions, but in this case, the substances which give only half the molecular lowering of the normal, are either non-electrolytes like the alcohols, or weakly dissociated acids like formic, acetic, benzoic, etc. The probable significance of the small molecular lowering produced by some substances is, that they are in a state of molecular aggregation in the particular solvent. When the molecular lowering, in the case of two undissociated compounds dissolved in a given solvent, is twice as great for one as for the other, it means that twice as many molecules of the second are aggregated into a unit, as of the first. If the molecules of the one exist singly in solution, those of the second are combined in twos. This will be seen at once, if we remember that the lowering of the freezing-point of a solvent depends only on the relative number of parts of the solvent and of the dissolved substance.

When we come to the results with water as a solvent, we have to deal with an entirely new set of phenomena. The results given above are a few taken from a large number. Compounds like the non-electrolytes, give a molecular lowering for water, which is very nearly a constant, and which is approximately 18.8. This is true for such a large number[1] of substances which have

[1] Ann. chim. phys. [5], 28, 137.

been investigated that there is no reason for regarding them as being the exceptions. On the other hand, all the strong electrolytes, including the strong acids, strong bases, and salts of strong acids with strong bases, weak acids with strong bases, and weak bases with strong acids, give molecular lowerings which are greater than the value 18.8. The explanation which has been offered to account for this and related facts by Arrhenius,[1] is that the molecules of the electrolytes do not exist as such in water solution. They are dissociated into parts called ions, and the amount of such dissociation depends, for a given substance, chiefly upon the amount of water present—on the dilution of the solution. In a very dilute solution of a strongly dissociated electrolyte, we have practically no molecules present, only ions. If the molecule is binary, each yields two ions, and since an ion lowers the freezing-point as much as a molecule, the molecular lowering for such substances, at high dilution, is twice as great as where there is no dissociation. If the molecule dissociates into three ions, and the dilution is such that the dissociation is complete, the lowering of the freezing-point will be three times as great as where there is no dissociation as with the non-electrolytes.

It is stated above that in aqueous solutions the molecules break down into parts called ions. It is so easy to confuse ions with atoms, and this is so frequently done, that a word of caution here is hardly out of place. An ion is not an atom, but is <u>an atom</u> charged with electricity. The resemblance between the two is far less close than might be imagined, except in weight. The properties of many of the atoms could not be foretold from the properties of the ions, with any de-

[1] Ztschr. phys. Chem., 1, 631.

gree of probability. An atom of potassium has properties so different from an ion of potassium, that one is more impressed by their difference than by their resemblance.

In some cases, as with the non-electrolytes, we have then to deal only with molecules in solutions in water, while with the electrolytes we have both molecules and ions, or only ions, depending on the dilution of the solution.

That the true value of the molecular lowering for water, when there is neither molecular aggregation nor electrolytic dissociation, is 18.8, has been shown theoretically by van't Hoff,[1] and more clearly presented by Ostwald,[2] thus:

Given a solution which contains n molecules of the dissolved substance and N molecules of the solvent. Let T be the temperature of solidification of the solvent and Δ the lowering of its freezing-point. Let enough of the solvent solidify to dissolve a molecule of the substance, $\left(\dfrac{N}{n} \text{ molecules}\right)$. Let λ be the molecular heat of fusion of the solvent, the amount of heat set free $= \dfrac{N}{n}\lambda$. If the ice is now separated from the solution, warmed to the temperature T, fused, and finally allowed to mix with the solution which has also been warmed to the same temperature, by passing through a semi-permeable membrane, an osmotic pressure p will be exerted. If the volume of the solvent which solidified is v, the work equals pv, the heat $\dfrac{N}{n}\lambda$,

$$\dfrac{pv}{\dfrac{N}{n}\lambda} = \dfrac{\Delta}{T} \quad \text{or} \quad \dfrac{pvn}{N\lambda} = \dfrac{\Delta}{T}. \quad \left(i.e.\ \dfrac{W}{Q} = \dfrac{\Delta}{T}\right)$$

[1] Ztschr. phys. Chem., 1, 481.
[2] Lehrbuch allgem. Chem., 1, 760.

But $pv = RT$ and $R = 2$ cal. Substituting we have:

$$\Delta = \frac{n}{N} \cdot \frac{2T^2}{\lambda}$$

Let M be the molecular weight of the solvent, and placing $N = \frac{100}{M}$, we have:

$$\Delta = \frac{nM}{100} \cdot \frac{2T^2}{\lambda} \quad \ldots\ldots\ldots\ldots\ldots (1)$$

In the Raoult formula $m = \frac{C}{A}$, m is the molecular weight of the dissolved substance, A is the specific lowering of the freezing-point, which equals $\frac{\Delta}{p}$, in which p is the percentage composition of the solution, and C is a constant. n, the number of molecules of the dissolved substance in 100 grams of the solvent $= \frac{p}{m}$.

$$m = \frac{Cp}{\Delta}, \quad m\Delta = Cmn.$$

$$\Delta = Cn \quad \ldots\ldots\ldots\ldots\ldots\ldots (2)$$

From (1) and (2) we have:

$$C = \frac{M}{100} \cdot \frac{2T^2}{\lambda}.$$

If L is the latent heat of fusion of a gram of the solvent, $\lambda = LM$, and

$$C = \frac{2T^2}{100L}.$$

The absolute temperature T, for the freezing-point of water, is 273°, and L, the latent heat of fusion of a gram of water, was taken by van't Hoff as 79°.[1] When these

[1] Ztschr. phys. Chem., 1, 497.

values are inserted in the above expression, $C = 18.9$. The value of L is probably more nearly 79.7 when C becomes 18.8.

I have shown experimentally[1] that the value of C for water, as determined with solutions of urea, ethyl and propyl alcohols, is respectively:

$$18.88$$
$$18.76$$
$$18.77$$

The formula of van't Hoff applies to the calculation of the constant for any solvent.

The freezing-point method has thus two distinct applications: To determine the molecular weight of compounds in solution, which are not dissociated by the solvent; and to measure the amount of the dissociation of electrolytes in solutions of different concentrations.

The applicability of the freezing-point method to the determination of the molecular weights of substances in solution, was pointed out by Raoult.[2] If we represent the unknown molecular weight of a substance by m, the molecular lowering or constant for the solvent by C, and the lowering of one per cent. of the dissolved substance by S, we have

$$m = \frac{C}{S}.$$

If the weight of the solvent used is W, that of the dissolved substance w, and the observed lowering of the freezing-point Δ,

$$S = \frac{\Delta W}{100\, w}.$$

---

[1] *Ztschr. phys. Chem.*, 12, 653.
[2] Compt. rend., 101, 1056.

Substituting this value of S in the above expression, it becomes

$$m = \frac{100\ Cw}{\Delta W}.$$

If the constant C is multiplied by 100 and termed C', the expression becomes

$$m = \frac{C'w}{\Delta W}.$$

The values of C' for a number of the solvents most commonly used are

	C'.
Water	1880
Benzene	4900
Phenol	7500
Formic acid	2770
Acetic acid	3880
Nitrobenzene	7070

The Application of the Freezing-Point Method to the Determination of Molecular Weights in Solution

Beckmann[1] has devised a form of apparatus which is both simple and efficient. C (Fig. 1) is a small glass battery-jar covered with some poorly conducting substance, and which is filled with the freezing material. A mixture of finely powdered ice and salt is convenient. B is a thick-walled glass tube, into which tube A, containing the solution, is inserted. A side tube attached to tube A, is thought to be useful in introducing the substance whose molecular weight is to be ascertained, but can readily be dispensed with. The thermometer, of the Beckmann differential type, is fitted into the tube A, by means of a cork, which can be easily removed. The stirrer S passes through the same cork, and must

[1] Ztschr. phys. Chem., 2, 638.

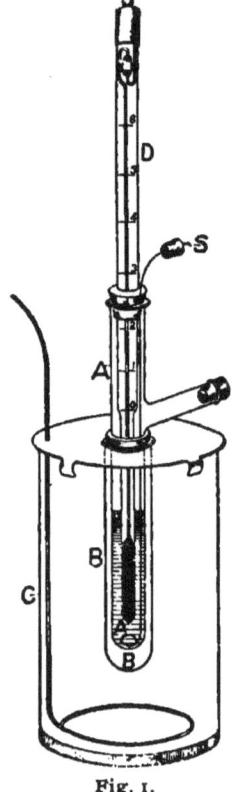
Fig. 1.

be of such form and dimensions as to move freely up and down between the inner walls of the tube and the bulb of the thermometer.

A small glass rod, bent at the bottom in the form of a ring, which will easily enter the glass tube A, is quite efficient. A short piece of glass tubing, through which this rod will move freely, is forced through a hole in the cork at the top of tube A, and serves both to hold the stirrer in place, and to allow smoother movement through the cork. The apparatus of the following dimensions has been found in this laboratory to be convenient.

Tube A is 20 cm. in length and 3 cm. in width. B is about 15 cm. long and 5 cm. wide. The glass tube used in constructing the stirrer should be about 2.5 mm. in thickness. A thermometer with a short, thick bulb, such as is sometimes furnished on the market, is not as desirable as one whose bulb is longer and of smaller diameter, since it requires a longer time to register the temperature of the liquid.

In case the solvent used is hydroscopic, some precaution must be taken to protect it from the moisture in the air. An apparatus, satisfying this requirement, has been constructed also by Beckmann,[1] by forcing the air which enters the apparatus to pass over some drying agent, like sulphuric acid. The device is shown in

[1] Ztschr. phys. Chem., 7, 324.

Fig. 2. The handle of the stirrer E passes through a glass tube, into which the side tube F, containing a few drops of sulphuric acid, is fused. The air enters through this side tube, is dried, and passes out through the tube receiving the handle of the stirrer. The remainder of the apparatus is of exactly the same form as shown in Fig. 1, except that it is provided with a glass siphon H, for removing the melted freezing-mixture. This is really superfluous, since a piece of rubber tubing answers the purpose equally well.

Forms of apparatus far more accurate than those

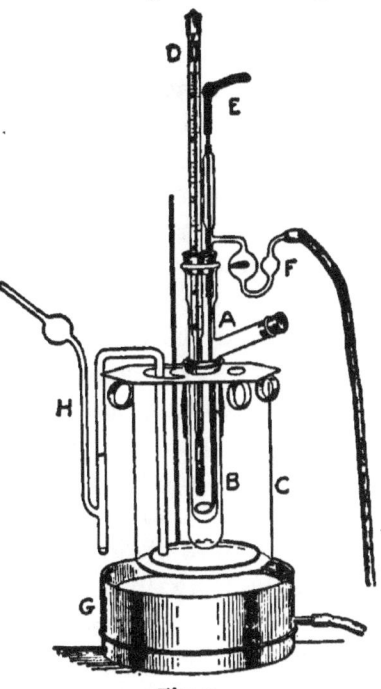

Fig. 2.

just described, have been devised and used, but since such extra refinement is desirable rather to measure dissociation than to determine molecular weights, reference will be given to them under the second application of the freezing-point method.

Carrying Out a Determination

The thermometer must first be so adjusted that the freezing-point of water falls near the top of the scale. To accomplish this, water is poured into the tube A, until the bulb of the thermometer, when placed in position, is covered. Tube A is placed directly in the freezing-mixture in C, and the water allowed to freeze. As soon as fine particles of ice separate, tube A is removed from the freezing-mixture, placed in tube B, and the whole then placed again in the freezing-mixture. The thermometer is then raised out of the water containing ice particles, allowed to remain in contact with the warmer air a moment, and then given a sudden jar. The mercury falls from the top to the bottom of the upper cup, and leaves the column free at its upper end. The thermometer is then placed again in the ice-cold water, and if the end of the mercury column does not come to rest on the upper half of the scale, the process just described is repeated. A few trials generally suffice to bring the reading approximately where desired.

The thermometer being adjusted, tube A is carefully dried, closed at the top and side with wooden stoppers, and weighed. Enough pure water is poured into the tube to cover the bulb of the thermometer when in position, and the tube is again weighed. The weight of the solvent employed is thus determined. The stopper is then removed from the top of the tube and the thermometer and stirrer placed in position. Tube A is placed in tube B, and the whole system in the freezing-mixture. During the cooling of the solvent the stirrer should be raised and lowered frequently. The water will cool down below its freezing-temperature often a degree or more, before the ice will begin to separate. When the undercooling of the solvent or of a solution is very much more than a degree, a small fragment of pure ice should be thrown into the overcooled liquid. This will start the separation of ice, which will continue until the true freezing-temperature is reached.

When the ice begins to separate, the mercury column will rise, rapidly at first, then slower, until it reaches the point of equilibrium. While the thermometer is rising, and especially when near the point of rest, it must be tapped gently to prevent the mercury from lagging back in the capillary, due to friction against its walls. A lead pencil is convenient to use in jarring the thermometer. The freezing-point of the water is then noted on the thermometer. The reading on the ordinary Beckmann instrument can easily be made to $0.001°$ by means of a small pocket lens.

The tube containing the solvent, with the thermometer and stirrer in position, is removed from the freezing-mixture, and the ice melted, by seizing the tube for a few moments with the hand. The freezing-point of the water is then redetermined exactly as described above. The two determinations should not differ more than two- or three-thousandths of a degree.

The substance whose molecular weight is to be determined, is weighed in a weighing tube, poured into the solvent, and brought completely into solution. If cane-sugar is used, that quantity is taken which will give a solution about one-tenth normal. If urea, or any of the alcohols is used, a more concentrated solution may be employed. A solution of cane-sugar, dextrose, etc., more concentrated than one-tenth normal, gives abnormally large depressions of the freezing-point of water. The reason for this is not entirely clear. The solution is then placed in the freezing-mixture and its freezing-point determined, and redetermined, exactly as described for the solvent. All the data are thus available for calculating the molecular weight of the substance from the expression already given.

Correction for the Separation of Ice

A certain amount of the solvent separates in the solid form in all such determinations, and the solution becomes concentrated by just this amount. The freezing-point of the solution, as read on the thermometer, is therefore always lower than would correspond to a solution of the concentration originally used. A correction for the change in concentration, due to the separation of the solid solvent, must be introduced. The amount of the solvent which separates in the solid phase, can easily be determined, knowing the amount by which the solution is undercooled before the ice begins to separate, the latent heat of fusion of a unit quantity of the solvent, and the specific heat of the liquid. The fraction of the solvent which separates is calculated thus, as was pointed out by the present writer:[1]

If we represent by u the amount of the undercooling of the solution in degrees centigrade, by w the latent heat of fusion of unit weight of the solvent, by s the specific heat of the liquid, and by T the fraction which will solidify, we have

$$T = \frac{su}{w}.$$

When water is used as a solvent $s = 1$, and $w = 80$. The fraction of this solvent which will separate as a solid, for every degree of undercooling, is therefore $\frac{1}{80}$, and the concentration of the original solution is increased by just so much. Instead of applying the correction to the concentration, it is simpler to apply it directly to the freezing-point lowering itself.

The Application of the Freezing-Point Method to the Measurement of Electrolytic Dissociation

An ion lowers the freezing-point of a solvent just as much as a molecule. If a molecule dissociates into two ions it will lower the freezing-point of a given amount of

---
[1] Ztschr. phys. Chem., 12, 624.

THE FREEZING-POINT METHOD 15

a solvent, just twice as much as if it is not dissociated. The lowering of the freezing-point of a given solvent by a partially dissociated electrolyte, depends upon the relation between the number of molecules of the solvent, and the sum of the molecules plus the ions of the dissolved substance. Thus, it is possible for any given dilution, to determine the amount to which an electrolyte is dissociated. The calculation of the dissociation from the freezing-point lowering is simple. The molecular lowering of the freezing-point of any solvent by any substance was defined by Arrhenius[1] as the lowering produced by a gram-molecular weight of the substance in a liter of solution. This can be taken as approximately one-tenth of the molecular lowering as defined by Raoult. For our present purpose we accept the definition of Arrhenius, and find that the molecular lowering of water produced by a gram-molecular weight of a non-electrolyte, like urea, the alcohols, etc., in a liter of solution, is the constant 1.88°. If the substance used is dissociated, the molecular lowering is always greater than 1.88°. The first step is to calculate the molecular lowering for the solution in question, which is done by dividing the lowering found, by the concentration in decimal part of normal. If there were only molecules present the molecular lowering would be 1.88°. The molecular lowering found must therefore be divided by 1.88, which gives the value of the van't Hoff coefficient i, for the solution.[2]

$$\frac{Molecular\ lowering}{1.88} = i.$$

If the molecule breaks down into two ions, the percent-

[1] Ztschr. phys. Chem., **2**, 494.
[2] *Ibid*, **1**, 501.

age of dissociation, α^1 (Arrhenius activity coefficient), is expressed thus

$$\alpha = i - 1.$$

If the molecule breaks down into three ions,

$$\alpha = \frac{i-1}{2}.$$

If into n ions,

$$\alpha = \frac{i-1}{n-1}.$$

The Method of Work

Exactly the same apparatus may be used as was employed in the determination of molecular weights. The method of preparing the solutions is, however, somewhat different.

The solvent is poured into the innermost vessel in quantity large enough to cover the bulb of the thermometer, and its freezing-point upon the thermometer ascertained, as in a molecular weight determination. The solvent is then completely removed from the vessel and the solution of known concentration, prepared in a measuring flask, introduced. Its freezing-point is then determined exactly as previously described, including the rapid stirring, the tapping of the thermometer, the introduction of a fragment of the solid solvent when necessary, and the correction for the change in concentration due to the separation of the solid solvent. The dilution of the solution is then increased one and a half, two, three, four times, etc., and the dissociation determined for each dilution. It will be found that the value of i, and therefore of α, always increases with increase in dilution. In

[1] Ztschr. phys. Chem., 11, 535.

this work any of the common chlorides, nitrates, bromides, or in general any electrolyte may be used.

It is convenient to use a solution of pure sodium or potassium chloride of concentration about 0.5 normal, and then to increase the dilution of this solution in several steps, as indicated above. The chlorides and nitrates break down into two ions each, the sulphates into three. The values of α from the freezing-point method should be preserved and compared with the values of α for the same solutions, as obtained by the conductivity method.

Far more accurate experimental methods have been devised and used for measuring the freezing-point lowerings, by Loomis,[1] Nernst and Abegg,[2] Ponsot,[3] myself,[4] and others.

A form of apparatus, which was found by the writer to give excellent results, is sketched in Fig. 3.

A is a large metallic vessel, 25 cm. high and 35 cm. wide. This is surrounded by a mantle of non-conducting material to protect it from the warmer air. B is a vessel of zinc, 21 cm. high and 15 cm. wide, which rests upon a tripod, to diminish the surface of contact with the outer vessel. This is provided with a lid of zinc. The vessel B is completely surrounded, except above, with a freezing-mixture of ice and a little salt. The space between A and B, filled with the freezing-mixture, was covered with the ring of asbestos, aa, to protect the freezing-mixture from the air. C is a glass vessel, 18 cm. high, 10 cm. wide, and of about 1200 cc. capacity. This rests on a thick felt bottom, which protects it from the zinc vessel beneath. The space between B and C is

---

[1] Ber. d. chem. Ges., **26**, 797; Wied. Annalen, **51**, 500.
[2] Ztschr. phys. Chem., **15**, 681.
[3] Compt. rend., **122**, 668.
[4] Ztschr. phys. Chem., **11**, 110, 529.

filled with air and covered above with a ring of felt, bb, which rests on a metallic shelf fastened on to the inner side of the vessel B. The air-chamber between B and C is thus closed and remains at nearly the same temperature during a determination. The glass vessel was covered with a glass lid. D is a thermometer

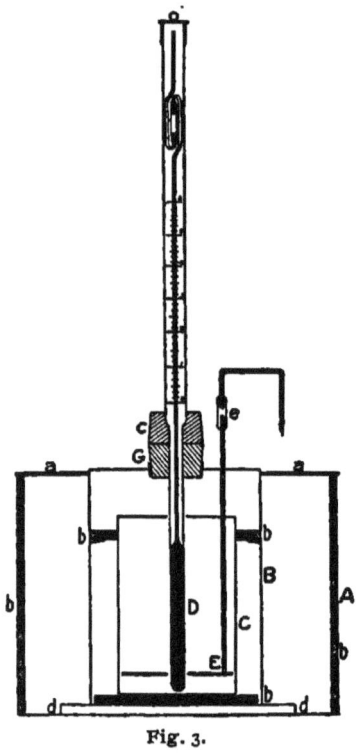

Fig. 3.

whose bulb is 14 cm. in length and 1.5 cm. in width. The fine capillary was carefully calibrated. The entire scale, which was 22 cm. in length, corresponded to only 0.6°. It was divided into tenths, hundredths, and thousandths of a degree. The finest divisions could be estimated to tenths, through a telescope, so that the scale could be

THE FREEZING-POINT METHOD

read to 0.0001 of a degree. The thermometer was of the Beckmann type, and the freezing-point of the solvent could be adjusted upon the scale, wherever desired. It was fastened firmly in cork c, and passed loosely through g, being suspended in the liquid in C.

E is a stirrer, which was constructed as follows: A circular piece of sheet-silver was cut somewhat smaller than the glass vessel, and plated electrolytically with gold. This was cut along the circular lines shown in Fig. 4, and also horizontally, as shown. The ends

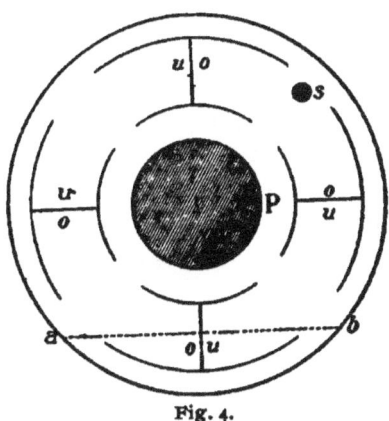

Fig. 4.

marked o were bent upwards, those marked u downwards. S is a small hole which received the handle. P is a large hole in the center, through which the bulb of the thermometer passed. In Fig. 5 is given a section

Fig. 5.

across one of the openings, to show how the ends are cut and bent. This section corresponds to the dotted line a,b in Fig. 4.

A stirrer of this form has the advantage that at every

movement up and down the liquid is moved horizontally and vertically, and any currents set up during the stroke in one direction are completely reversed by the opposite stroke. The advantage claimed for this method is that by using a large volume of the solution the temperature can be much better regulated. The comparatively thick layer of air at constant temperature, around the innermost vessel, makes it far less susceptible to the influence of changes in the temperature of surrounding objects. The large volume of the liquid exposes relatively less surface to the cooling mixture, and the rate of cooling is comparatively slow. This makes it possible to determine more accurately the temperature of the liquid in which the ice separates. Since the rate of cooling is slow, the ice which separates during the time required for the thermometer to become constant, is relatively small.

A liter of pure water is placed in the vessel C, and its freezing-point determined on the thermometer. A certain volume of this is then removed, and an equal volume of a solution of known concentration, added. Thus the volume of the solution in the vessel always remains a liter, which facilitates the calculation of the results. The same process is repeated in making successive dilutions. By this method of work the first solution of a series is the most dilute, and these become more and more concentrated to the end of the series.

Such accuracy as is reached with this apparatus is not absolutely necessary for laboratory practice, but is very desirable where the problem of the measurement of electrolytic dissociation presents itself.

The applicability of the method will be seen by comparing the values of the dissociation of a number of acids, bases, and salts, as obtained by it,[1] with the dissociation as determined by the conductivity method of Kohlrausch.[2]

Substance.	Concentration normal.	Dissociation from conductivity. Per cent.	Dissociation from lowering of freezing-point. Per cent.
NaCl	0.001	98.0	98.4
	0.010	93.5	90.7
	0.100	84.1	83.5
K_2SO_4	0.002	92.2	94.1
	0.010	85.8	88.2
	0.100	70.1	72.0
$BaCl_2$	0.002	93.9	94.1
	0.010	87.9	88.4
	0.100	75.3	76.8
HCl	0.002	100.0	98.4
	0.010	98.9	95.8
	0.100	93.9	88.6
H_2SO_4	0.003	89.8	86.0
	0.005	85.4	83.8
	0.050	62.3	60.7
HNO_3	0.002	100.0	98.4
	0.010	98.5	96.8
	0.100	93.5	87.8
H_3PO_4	0.002	87.8	85.2
	0.010	63.5	68.8
KOH	0.002	100.0	98.4
	0.010	99.2	93.7
	0.100	92.8	83.1
NaOH	0.002	98.9	98.4
	0.010	99.5	93.7
	0.050	90.4	88.4

An absolute agreement between the dissociation values obtained by the conductivity method and by the

[1] Ztschr. phys. Chem., 12, 639.
[2] Wied. Annalen, 26, 161.

freezing-point method, is not to be expected, since the former method was used at 18° or 25°, while the latter was applied at about 0°.

PART II

THE BOILING-POINT METHOD

Theoretical

The presence of a foreign non-volatile substance diminishes the vapor pressure of the solvent in which it is dissolved. Since the boiling-point of a solvent, or of a solution, is the temperature at which the vapor-pressure just overcomes the pressure of the atmosphere, it follows that the solution having a lower vapor-pressure than the solvent, will have a higher boiling-point.

There are thus two quantities, either of which we may measure: the depression of the vapor-tension of the solvent, caused by the presence of the dissolved substance; or the rise in the boiling-point of the solvent, due to the same cause.

Passing over the work of Faraday,[1] Griffiths,[2] Legrand,[3] and others, along this line, since it all fell short of any very important generalization, we come to that of von Babo,[4] who found that the relation between the amount of salt present and the diminution of the vapor-pressure, was independent of the temperature.

The work of Wüllner[5] was of greater significance. He measured the depression of the vapor-pressure of water especially by salts, and arrived at the conclusion that the diminution of the vapor-pressure of water, pro-

[1] Ann. chim. phys., 20, 324.
[2] Pogg. Annalen, 2, 227.
[3] Ann. chim. phys., 59, 423.
[4] Jahrb. Chem., 1848-49, 93; 1857, 72.
[5] Pogg. Annalen, 103, 529; 105, 85.

duced by dissolved non-volatile substances, was proportional to the amount of substance present.

While this is true only in certain cases, or indeed only for certain classes of compounds, yet it is strictly analogous to the earliest generalization reached in connection with the study of the depression of the freezing-point of a solvent by a foreign substance. It will be remembered that Blagden stated that the depression of the freezing-point of a solvent by a dissolved substance was proportional to the amount of substance present.

When depressions of the freezing-point were measured with a fair degree of accuracy, it was shown that the generalization of Blagden held only approximately, and in some cases. So also, the relation pointed out by Wüllner was shown to be only an approximation under certain conditions, by the work of Pauchon,[1] Tammann[2] and Emden.[3]

It is to Raoult, more than to any other, that we owe the theoretical development of the subject in hand.

He employed the first of the two quantities mentioned, and measured the depression of the vapor-tension of a solvent by foreign substances. A number of relations were brought out by him from a study of solutions in solvents other than water, which would not have been discovered in aqueous solutions, since these are often dissociated, and to a different amount for different dilutions.

Raoult[4] confirmed the generalization of von Babo, that the relation between the depression of the vapor-pressure and the vapor-pressure of the solvent, was independent of the temperature between 0° and 20°. Also that of Wüllner, that the depression of the vapor-pres-

[1] Compt. rend., 89, 752.
[2] Wied. Annalen, 24, 523.
[3] *Ibid*, 31, 145.
[4] Compt. rend., 103, 1125.

sure was proportional to the concentration (when there was no dissociation).

If we represent the vapor-pressure of the pure solvent by p, and that of the solution by p',

$$\frac{p-p'}{p}$$

is independent of the temperature and proportional to the concentration.

The nature of the substance used was then investigated to determine whether the chemical composition of the molecule had any effect on its power to depress the vapor-tension.

Solutions containing the same number of molecules of different substances in solution in the same number of molecules of a given solvent, must be compared as to their vapor-pressures. This would be difficult to carry out directly. Convenient concentrations of different substances were used, the vapor-pressures of the solutions determined, and the molecular depression of the vapor-pressure calculated for each substance from the expression

$$\frac{p-p'}{p} \times \frac{m}{g},$$

in which m is the molecular weight of the substance, and g the number of grams in 100 grams of the solvent.

He found that the molecular depression of the vapor-pressure of a solvent is a constant for a given solvent, independent of the nature of the substance which is dissolved in it. This holds, as we now know, only when the dissolved substances are <u>undissociated</u> by the solvent.

Raoult[1] investigated also the relative diminution of the vapor-pressure of different solvents, when the rela-

[1] Compt. rend., 104, 1430.

tion between the number of molecules of the substance and that of the solvent was the same.

Below are given the results of the relative diminution of the vapor-pressure for twelve solvents, calculated on the basis of one molecule of the substance to 100 molecules of the solvent:

Water	0.0102
Phosphorus trichloride	0.0108
Carbon bisulphide	0.0105
Tetrachlormethane	0.0105
Chloroform	0.0109
Amylene	0.0106
Benzene	0.0106
Methyl iodide	0.0105
Methyl bromide	0.0109
Ether	0.0096
Acetone	0.0101
Methyl alcohol	0.0103

The relative diminution of the vapor-pressure of solvents produced by a' molecule of a non-volatile substance, in the same number of molecules of the solvents, is very nearly a constant. This relation has been satisfactorily formulated thus:

$$\frac{p-p'}{p} = c \cdot \frac{n}{N+n},$$

in which n is the number of molecules of the dissolved substance, N that of the solvent, and c a constant, which, from the foregoing table, can be regarded as unity. The expression becomes then

$$\frac{p-p'}{p} = \frac{n}{N+n},$$

or the lowering of the vapor-pressure of the solvent is to the vapor-pressure of the solvent, as the number of molecules of the dissolved substance is to the entire

THE BOILING-POINT METHOD

number of molecules present. This expression was tested experimentally by Raoult,[1] and found to hold for a large number of substances in ethereal solution.

From the foregoing it will be seen that the molecular weight of substances can be determined directly from the depression of the vapor-pressure of a solvent which they produce.

Let the molecular weight of the substance be represented by m, and the amount used by a, then the number of molecules, $n = \dfrac{a}{m}$.

Making $N = 1$, and substituting this value of n in the expression

$$\frac{p-p'}{p} = \frac{n}{N+n},$$

we have

$$m(p-p') = p'a,$$

$$m = \frac{p'a}{(p-p')}.$$

Knowing p, p' and a, we calculate m directly.

As a practical method for determining molecular weights, this is not used, since such measurements are not easily carried out, and are not very accurate.

It has been found to be simpler and more accurate to determine the temperature at which the vapor-pressure of the solution is equal to that of the solvent. Since the boiling-point of a liquid is the temperature at which its vapor-pressure just overcomes the pressure of the atmosphere, the boiling-points of a solvent and of a solution in that solvent, are the temperatures of equal vapor-pressures. The object is then to determine the rise in the

[1] Ztschr. phys. Chem., 2, 371.

boiling-point of a solvent produced by the substance dissolved in it.

The experimental method for carrying out such determinations we owe to Beckmann. The forms of apparatus which seem best adapted to this work, and the details of an experiment will be considered in the second part of this chapter.

The rise in the boiling-point is directly proportional to the lowering of the vapor-pressure, and depends upon the relative number of molecules of the solvent and of the dissolved substance.

The probability of calculating molecular weights directly from the rise in the boiling-point of solvents produced by substances dissolved in them, is at once apparent.

The expression by which the molecular weights are calculated, is analogous to that already given for calculating molecular weights from lowerings of the freezing-point of solvents by dissolved substances. If we represent the unknown molecular weight by m, the weight of the substance used by w, the weight of the solvent by W, and the rise in the boiling-point of the solvent by R, we have

$$m = \frac{Cw}{RW},$$

in which C is a constant of different value for each solvent.

The analogy between the lowering of the freezing-point and the rise in the boiling-point holds still further, in that the value of C can be calculated from the same formula:

$$C = \frac{2T^2}{100\,L}.$$

When applied to calculating the constant for the boiling-point method of determining molecular weights, T is the absolute temperature at which the pure solvent boils, and L the latent heat of evaporation of the solvent.

The values of C for a number of the solvents more commonly used, are given below. The solvents are arranged in the order of their boiling-points.

	C.	Boiling-point.
Ethyl ether	2110	34.9
Carbon bisulphide	2370	46.0
Acetone	1670	56.3
Chloroform	3660	60.2
Ethyl alcohol	1150	78.3
Benzene	2770	79.6
Water	520	100.0
Acetic acid	2530	118.0
Ethylene bromide	6320	128.8
Aniline	3220	184.5

Some of the relations between the boiling-point and the freezing-point methods have been mentioned, but others, however, exist. It will be remembered that the freezing-point method can be used to determine the molecular weights of only a limited number of substances—the non-electrolytes. The boiling-point method is subject to the same limitation. Those substances which give abnormally great depressions of the freezing-point, due to electrolytic dissociation, give abnormally great depressions of the vapor-tension, or rise in the boiling-point. It was pointed out that in such cases the freezing-point method could be used to measure the amount of the dissociation in the solutions. The boiling-point method may be used for the same purpose, but is not capable of the same degree of accuracy as the former.

Some recent work[1] has shown, however, that it can be applied to the problem of electrolytic dissociation in solution, and when all the precautions specified are taken, it is capable of giving fairly satisfactory results.

The Application of the Boiling-Point Method to the Determination of Molecular Weights in Solution

The precautions which are necessary in making such measurements will be understood best by pointing out the more prominent errors to which the boiling-point method is subject.

The method as such, is not capable of that refinement to which the freezing-point method has been developed. It is sensitive to barometric changes, which seriously affect the boiling-point of liquids. In this method the vapor escapes quickly from the solution in which its presence is necessary to establish the temperature equilibrium. The difference in temperature between the liquid and surrounding objects is generally much greater in this method than in the freezing-point method, so that more precautions are necessary to protect the solution and thermometer from changes in the temperature of external objects. In this method a part of the comparatively pure solvent is constantly separating from the solution as vapor, and is returned as a liquid, at a temperature lower than that of the boiling solution. The amount of this liquid cannot, for given conditions, be determined with the same degree of accuracy as was possible in ascertaining the amount of ice which separated in the freezing liquid. The large Beckmann thermometers are more liable to undergo change at the comparatively high temperatures to which they are subjected in this method, than in the

[1] Jones and King: Am. Chem. J., **19**, 581.

freezing-method, where the temperature of the thermometer is at no time widely removed from the ordinary. The boiling-point method has this advantage that more solvents can be employed, since comparatively few solvents freeze within the range of ordinary temperatures; and further, the solubility of substances is generally increased at the higher temperatures. It is, on the other hand, a misfortune for the boiling-point method that aqueous solutions cannot be used satisfactorily, partly because of the very small constant for water.

A number of forms of apparatus for determining the boiling-point of solvents and solutions have been devised by Beckmann,[1] to whom we are as much indebted for the experimental development of the subject as we are to Raoult for its theoretical. That one, which, judged by the results,[2] seems to be on the whole, the most satisfactory, is seen in Fig. 6. The inner glass vessel A, provided with a return condenser K, receives the liquid whose boiling-point is to be determined. The bottom of this vessel is filled to a depth of a few centimeters with glass beads or small garnets, so that the boiling may take place from a number of points and proceed more smoothly. The bulb of the Beckmann thermometer is placed well below the surface of the liquid. Tube A is surrounded on the sides with a double-walled glass jacket B, into which some of the same solvent placed in A is poured. The object is to surround the boiling liquid with a liquid as nearly at its own temperature as possible. This jacket is provided with a return condenser K_2. The whole is supported on a box

[1] Ztschr. phys. Chem., 4, 544; 8, 224; 15, 663; 21, 246.
[2] *Ibid*, 8, 224.

of asbestos C, which is open beneath. Heat is applied as shown in the drawing.

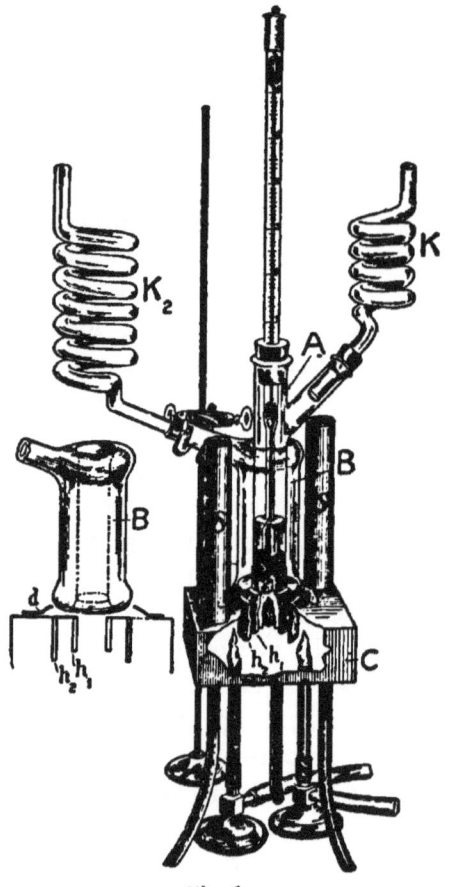

Fig. 6.

The results obtained by Beckmann[1] with the use of this apparatus were very good. It is, however, not free from objections. It is a question whether the effect of radiation from the bulb of the thermometer outward upon the colder objects, was entirely cut off by the form of

[1] Ztschr. phys. Chem., 8, 226.

jacket employed. Beckmann[1] used in a later form a porcelain jacket, having abandoned a metallic one, which doubtless cut off the radiation more effectively than the one of glass. But an objection which applies to every form devised by Beckmann, is that the cold solvent from the condenser is returned directly into the hot liquid in which the thermometer is immersed. That the thermometer is affected by this, in that it tends to lag behind the true boiling-temperature of the liquid, is probable.

A form of apparatus which largely eliminates this latter source of error, was devised by Hite,[2] and is shown in Fig. 7. The distinctive advance made by Hite is the introduction of an inner glass tube, which prevents the condensed solvent from coming in contact with the thermometer before it is reheated to the boiling-point. The cooled liquid must pass through a layer of the boiling liquid between the walls of the inner and outer vessel, some centimeters deep, before it can enter the inner tube which receives the thermometer. The inner vessel is closed at the bottom by means of a glass stopper. Grooves are filed into the edge of the stopper to allow the vapor to stream through into the inner vessel in fine bubbles, and stir the liquid around the thermometer. I am inclined to lay rather less stress upon the importance of this device, than upon the separation of the condensed solvent from the liquid in which the thermometer is placed, until it has been reheated to the boiling-point. The apparatus gave admirable results with low-boiling solvents, but could not be used for solvents which boil over 100°.

The present writer[3] has devised a form of apparatus

[1] Ztschr. phys. Chem., 15, 662.
[2] Am. Chem. J., 17, 514.
[3] *Ibid*, 19, 581.

which aims both at reducing the error from radiation to a minimum, and at preventing the condensed solvent from coming in contact with the thermometer until it is reheated to the boiling-point. It is also one of the simplest of the efficient forms thus far devised. It is shown in section in Fig. 8. A is a glass tube 18 cm. high and 4 cm. in diameter, drawn out at the top to a diameter of about $2\frac{3}{4}$ cm. and ground to receive a ground-glass stopper. This tube is filled to a depth of from 3 to 4 cm. with glass beads. P is a cylinder of platinum, 8 cm. high and $2\frac{1}{2}$ cm. in width, made by rolling up a piece of platinum foil, and fastening it in position by wrapping it near the top and bottom with platinum wire. Into the cylinder P, some pieces of platinum foil are thrown. These are made by cutting foil into pieces about $\frac{3}{4}$ cm. square, bending the corners alternately up and down, to prevent them from lying too closely upon one another, and serrating the edges with scissors, to give a greater number of points from which the boiling can take place. The bulb of the thermometer is thus entirely surrounded by metal at very nearly its own temperature, except directly above. A condenser C, about 40 cm. in length, is attached to the tube A_1, which is 2 or $2\frac{1}{4}$ cm. in diameter, by means of a cork. When it is desired to protect the solvent from the moisture in the air, the top of the condenser tube should be provided with a tube containing calcium chloride or phosphorus pentoxide. During an experiment, the vessel A is closed above by a cork, through which the Beckmann boiling-point thermometer T passes. M is a jacket of asbestos, 12 cm. high and $1\frac{1}{2}$ cm. thick, over the top of which the rate of boiling can be observed satisfactorily. It is constructed by bending a thin board of asbestos tightly around the tube A, and fixing it in place by means of a copper

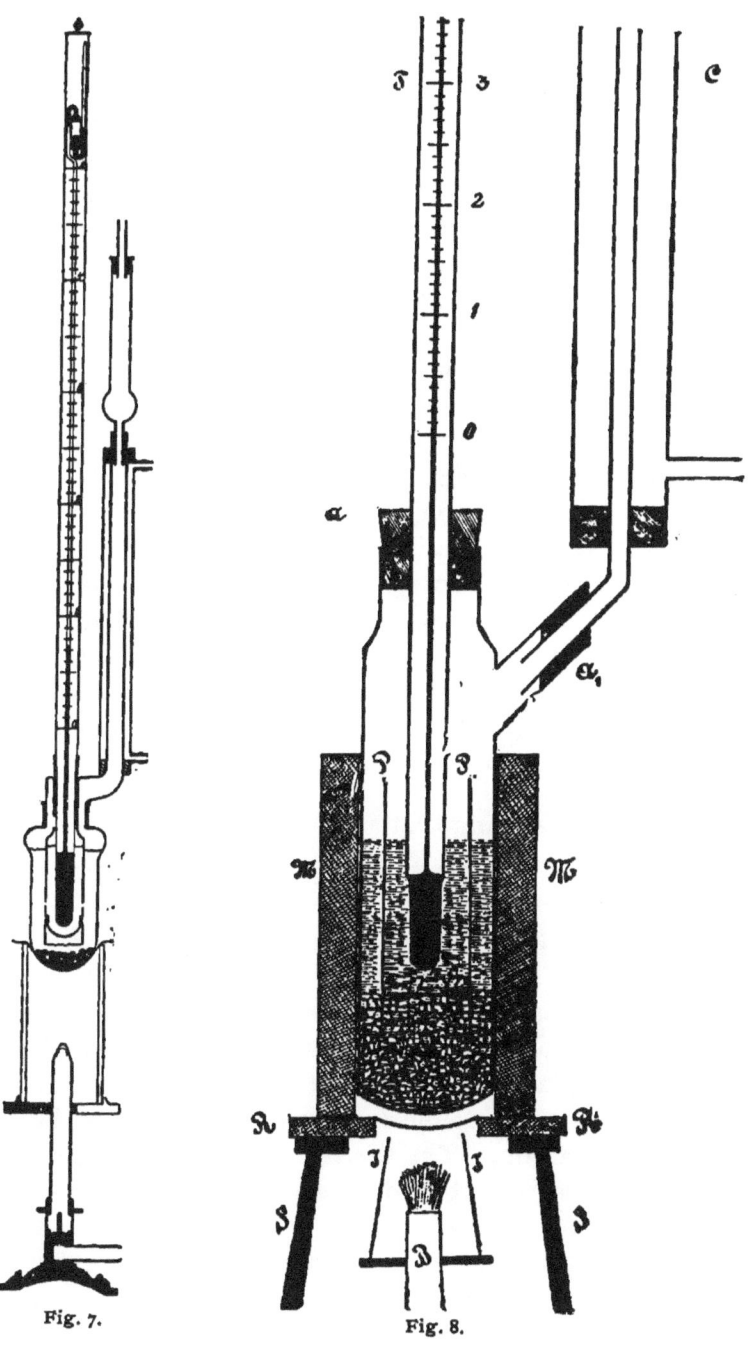

Fig. 7. Fig. 8.

wire. Thick asbestos paper is then wound around this until the desired thickness is reached. The apparatus is supported on a small iron tripod S, 8 cm. in diameter, on which rests an asbestos ring R, about 9 cm. in external diameter. A circular hole is cut in the center of this ring, about $3\frac{1}{2}$ cm. in diameter, and over this is placed a piece of fine copper gauze. The source of heat is a Bunsen burner B, surrounded by an ordinary metallic cone I, to protect the small flame from air-currents. The glass vessel A is shoved down until it comes in contact with the wire gauze. Under these conditions a very small flame suffices when low-boiling solvents are employed, and not a large flame is required when a solvent like aniline is used.

A number of other forms of apparatus have been constructed for determining the boiling-points of liquids, but these either do not eliminate error sufficiently for accurate work, or are so complex that they can scarcely hope to find general application in the laboratory for the purpose of determining molecular weights.

Carrying Out a Determination

The thermometer must first be so adjusted that the top of the mercury thread comes to rest on the lower half of the scale, when the bulb is immersed in the boiling solvent. This is accomplished by placing some glass beads in cylinder A, and adding the pure solvent until the bulb of the thermometer will be covered when inserted in place. The solvent is then boiled, and as much mercury as possible is driven out of the lower bulb into the upper cup. The thermometer is then removed from the liquid, inverted for a few moments, when still more of the mercury in the bulb will run down into the cup. The thermometer is then quickly brought into normal position and given a sudden tap, when the mercury will fall from the top to the bottom of the cup and leave the column free. The bulb is again placed in the boiling solvent, and if the thread comes to rest where desired, the apparatus is ready for a determination. If not, the process must be repeated until the desired end is reached, which, however, does not usually require any considerable expenditure of time.

When the thermometer is adjusted, it must be removed, and the apparatus and beads entirely freed from the liquid. The glass beads are then poured into the glass cylinder, the platinum cylinder inserted, and pressed down into the beads to a distance of from $\frac{1}{2}$ to 1 cm. The platinum plates are then introduced into the platinum cylinder, the end of the tube A, closed with a cork, and the ground-glass stopper inserted into A. The apparatus is then set into a small beaker glass and weighed, the solvent introduced, and the whole reweighed. (Great care must be taken that not enough solvent is employed to boil over from one side of the platinum cylinder to the other.) In case a laboratory is not provided with a balance capable of weighing accurately 200 or 300 grams, the solvent must be weighed directly and poured into the apparatus. This method of procedure, for low-boiling solvents, is necessarily less accurate, due to loss by evaporation.

After the solvent is weighed, the glass stopper is removed, and the thermometer, fitted tightly into a cork, is placed in position, as shown in the drawing. The apparatus is then placed upon the stand in the mantle of asbestos, the cork removed from A, and the condenser attached. Heat is then applied and the solvent boiled. The size of the flame must be so regulated by means of a screw pinch-cock, that the boiling is quite vigorous, but not so violent as to be of an irregular or explosive character. A quiet, but very active boiling is absolutely essential to the success of the experiment. The time required to establish the true temperature of equilibrium between the pure liquid solvent and its vapor, is much greater than in the case of a solution. This is strictly analogous to what is observed with the freezing-point method. Here, the time necessary to establish the temperature of the equilibrium between the solid and liquid phases of the pure solvent, is always much greater than for a solution. Before taking a reading on the Beckmann thermometer, it is always necessary to give it a few sharp taps with a lead pencil, and indeed this should be done occasionally while the mercury is rising, and especially when it is near the point of equilibrium. The use of an electric hammer to accomplish this object is an unnecessary complication. A small hand-lens, magnifying a half dozen times, is quite sufficient to use in making the readings. It is always best to redetermine the boiling-point of the solvent. After this point has been ascertained, a tube containing the substance pressed into pellets, whose molecular weight it is desired to determine, is weighed, and a convenient number of these poured into the solvent, either through the condenser, or directly through the tube A when the solvent is not too volatile, and has ceased to boil. The tube is then reweighed, and the amount of substance introduced, thus ascertained. The boiling-point of the solution is then determined.

The carrying out of a determination with a low-boiling solvent is a much easier process than with one boiling at a considerably higher temperature.

Thus: when anisol or aniline is employed, much care and some experience are necessary to determine the rate of boiling which must be adopted. If the boiling is too

THE BOILING-POINT METHOD 39

slow, the thermometer will never reach the temperature of equilibrium; if so rapid that it is irregular and explosive, the thermometer may rise above the true point, and then suddenly drop below it at the moment when a large amount of the vapor is set free. In a word, for high-boiling solvents, the rate of boiling must be as vigorous as possible, in order to proceed with perfect regularity.

In all such determinations the barometer must be carefully observed; but after the boiling-point of the solvent has been determined, that of the solution can be ascertained so quickly, that the changes in the barometer during this short interval are usually so slight that they are negligible. Whenever they are of appreciable value, a correction must be accordingly introduced.

A portion of the nearly pure solvent is constantly being evaporated from the solution, and condensed on the walls of the apparatus itself, and in the condenser. The solution is thus more concentrated than would be calculated from the amount of substance and of solvent used. A correction must be introduced for the amount of the solvent which separates from the solution, as was necessary for the freezing-point method. Unfortunately, we cannot determine the amount in the boiling-point method with even the same degree of accuracy as in the freezing-point method.

The amount of the solvent which exists as vapor and condensed liquid, is given by Ostwald[1] as 0.2 gram, and 0.35 gram for water. But this evidently holds only under a special set of conditions, and must be taken, in general, as only a rough approximation.

Thus all the data are at hand for calculating the

[1] Hand und Hilfsbuch zur Ausführung Physiko-Chemischer Messungen, p. 224.

molecular weight of the substance in the solvent used, from the formula already given (page 28):

$$m = \frac{Cw}{RW}.$$

Below are given a few of the results obtained with my apparatus for solvents boiling from 34.9° to 182.5°.

SOLVENT, ETHER: $k = 2110$; BOILING-POINT, 34.9° AT 760 mm.

Naphthalene, 128.

FIRST SERIES.

	Ether. Grams.	Naphthalene. Grams.	Rise in boiling-point.	Molecular weight.
1	57.573	1.2365	0.357°	126.9
2	57.573	2.5155	0.716°	128.8
3	57.573	3.8733	1.110°	127.9

Mean, 127.9.

SOLVENT, BENZENE: $k = 2670$; BOILING-POINT, 80.36° AT 760 mm.

Naphthalene, 128.

	Benzene. Grams.	Naphthalene. Grams.	Rise in boiling-point.	Molecular weight.
1	70.560	0.7594	0.215°	133.7
2	70.560	2.0548	0.574°	135.4
3	70.560	3.0780	0.850°	137.0
4	70.560	4.4790	1.234°	137.4

Mean, 135.9

SOLVENT, ANILINE: $k = 3220$; BOILING-POINT, 182.5° AT 738 mm.

Triphenylmethane, 244.

FIRST SERIES.

	Aniline. Grams.	Triphenylmethane. Grams.	Rise in boiling-point.	Molecular weight.
1	60.126	0.8017	0.180°	238.6
2	60.126	1.6052	0.353°	243.5
3	60.126	2.2914	0.496°	247.4
4	60.126	2.9213	0.654°	239.2

Mean, 242.2

THE BOILING-POINT METHOD 41

Diphenylamine, 169.

	Aniline. Grams.	Triphenylmethane. Grams.	Rise in boiling-point.	Molecular weight.
1	64.220	0.7780	0.224°	174.1
2	64.220	1.3326	0.391°	170.9
3	64.220	1.7832	0.535°	167.1

Mean, 170.7

For practice in the laboratory it is far better to use solvents with low boiling-points, such as ether or benzene. Ethyl alcohol can be employed, but with it the results are liable to be less accurate, since its constant is comparatively small.

Naphthalene is easily obtained pure, and may be used in both ether and benzene. In alcohol: benzoic acid, urea, or acetamide may be conveniently used; while triphenylmethane, diphenyl amine or benzanilide, give good results with aniline as a solvent.

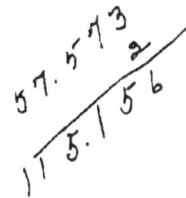

PART III

THE CONDUCTIVITY METHOD

Conductors of electricity may, for the sake of convenience, be divided into two classes, those which conduct without undergoing any decomposition, such as the metals, carbon, etc., and those which, during the passage of the current, undergo a decomposition or electrolysis at the poles, such as solutions of many substances. It is not at all certain that there is any very fundamental difference between the two classes, and at present, it seems that a resemblance between the two modes of conduction is becoming clearly recognized.

It is by no means true that solutions of all substances conduct. Thus, aqueous solutions of the so-called neutral organic compounds, such as the alcohols, carbohydrates, urea, and a large number of such substances, do not conduct the current. This furnishes ground for a division of substances into those whose solutions conduct and are called electrolytes, and those which, in solution, do not conduct and are called non-electrolytes.

The application of the conductivity method in physical chemistry is limited to conductors of the second class, $i.\,e.$, to solutions of electrolytes, which are chiefly solutions of acids, bases, and salts.

The conductivity of any conductor of electricity is the reciprocal of its resistance. The resistance r is, from Ohm's law, expressed thus:

$$r = \frac{\pi}{i}.$$

THE CONDUCTIVITY METHOD 43

π is the difference in potential at the two ends of the conductor, and i is the strength of current. The conductivity c is the reciprocal of r.

$$= \frac{i}{\pi}.$$

The unit of resistance, called the ohm, is that of a column of pure mercury 106.3 cm. long and 1 square mm. in section, at 0° C.

The Siemens or mercury unit is the resistance of a column of mercury 100 cm. in length and with a cross section of one square mm. The two units bear the relation to one another of 106.3 : 100.

Specific and Molecular Conductivities

The resistance of conductors depends upon their form as well as upon their chemical nature. In order that the resistance of different conductors should be measured in comparable quantities, their dimensions must be taken into account. The dimensions usually chosen are a cylinder 1 meter in length and 1 square mm. in section. The resistance of such forms of conductors, is known as their specific resistance. The reciprocal of this is their specific conductivity.

The conductors of the second class are solutions of some electrolyte in some solvent, and their conductivity depends chiefly or wholly upon the presence of the electrolytic substance. That the resistances of such solutions should be comparable, it is clear that we must deal with comparable quantities of the dissolved substances. The most convenient quantities are gram-molecular weights.

Given a normal solution which contains a gram-molecular weight of the electrolyte in a liter. If this liter of solution be placed between two electrodes which are 1 cm. apart, the cross section would be 1,000 square centimeters. This will have 0.001 of the resistance, or 1,000 times the conductivity of a cube of the same solution whose edge was 1 cm. in length. If we represent by v the number of cubic centimeters of any solution, which contains a gram-molecular weight of the dissolved substance, and by s the specific conductivity of a cube of the solution whose edge is 1 cm. in length, the molecular conductivity μ is the product of these quantities:

$$\mu = vs.$$

But if we represent by s the specific conductivity of a cylinder of the solution 1 meter in length and 1 square mm. in cross section:

$$\mu = 10,000\ vs.$$

A general expression, where g gram-molecular weights are contained in a liter of the solution, is:

$$\mu = \frac{s \times 10^3}{g},$$

when s, the specific conductivity, is referred to a cube of the solution, or:

$$\mu = \frac{s \times 10^7}{g},$$

when s is referred to a cylinder of the solution, 1 meter in length and a square mm. in cross section.

The molecular conductivities of solutions are then the conductivities of comparable quantities of different substances, and when the same dilutions are used, the molecular conductivities are directly comparable with one another.

Different substances behave very differently with respect to their power to carry the current, when in solution in a given solvent. The fundamental distinction between substances which conduct, and those which do not conduct at all, has been already mentioned. But among conductors very marked differences exist. Some reach a maximum of conductivity at moderate dilution, while others attain this only at extreme dilution. Take the case of a strong acid like hydrochloric or nitric; the molecular conductivity increases with the dilution to about one one-thousandth normal, when it becomes constant. While, on the other hand, the molecular conductivity of a weak acid like acetic, will increase with the dilution, as far as the dilution can be carried with the conductivity method.

The question arises, whence this difference between substances in respect to their power to carry the current? Here again, the theory of electrolytic dissociation comes to our aid. Those substances which give abnormally great depressions of the freezing-point, abnormally large elevations of the boiling-point, and which show abnormally great osmotic pressures, conduct the current; and only such substances conduct.

The explanation of the abnormal results with respect to the properties just mentioned, was sought in the dissociation of the molecules into ions. From a large amount of evidence from many sources, we seem justified in concluding that only ions conduct the current. Molecules are entirely incapable of carrying electricity through the solvent in which they are dissolved. If only ions conduct, then the conductivity of a solution is proportional to the number of ions present, provided that the ions move with the same average velocity, which is true of ions of the same kind.

The conductivity method can then be used to measure the dissociation of electrolytes in solution, and this is its most important scientific application. When the molecular conductivity attains a maximum constant value, it means that the dissociation is complete, and this value of the molecular conductivity is termed μ_∞, The molecular conductivity at any dilution is written μ_v. in which v is the volume of the solution, *i.e.*, the number of liters which contains a gram-molecular weight of the electrolyte. The percentage of dissociation at any dilution, α, is the ratio between the molecular conductivity at that dilution, and the molecular conductivity when the dissociation is complete:

$$\alpha = \frac{\mu_v}{\mu_\infty}.$$

In order to determine the dissociation of an electrolyte at any given dilution, by means of the conductivity method, it is necessary to determine the molecular conductivity, μ_v, at that dilution, and the value of μ_∞ for that electrolyte, when the value of α can be calculated at once.

Determination of μ_∞

The value of μ_v is determined directly for any electrolyte in any solvent, by means of the conductivity method. The determination of μ_∞ for strongly dissociated electrolytes is comparatively simple. The value of μ_v is determined at a given dilution, the dilution increased, the molecular conductivity determined at the new dilution, and this continued until a dilution is reached, which is so great, that when further increased, the value of μ_v remains the same. It has then attained a constant maximum value, which is the value of μ_∞. The value of μ_∞ for strong acids and bases, and for salts, is usually attained at a dilution between $v = 500$ and $v = 5000$. This will be seen from the following examples:

Hydrochloric acid.		Potassium hydroxide.		Potassium chloride.	
$v.$	μ_v 18°.	$v.$	μ_v 18°.	$v.$	μ_v 18°.
2	301	2	184.1	2	95.8
32	335	20	204.5	20	108.3
128	341	100	212.4	100	114.7
1000	346	500	214.0	1000	119.3
1667	344	1000	211.1	5000	120.9

A large number of substances, such as the organic acids and bases, which are only weakly dissociated at any ordinary dilution, present a new problem, when it is desired to determine their maximum molecular conductivity. That this is not reached at dilutions to which the conductivity method can be applied, is seen from the following examples:

Acetic acid.		Ammonia.[1]	
$v.$	μ_v 18°.	$v.$	μ_v 18°.
2	1.9	2	1.2
20	6.2	20	4.3
100	13.2	100	9.2
1000	38.0	1000	26.0
5000	79.6	5000	50.0
10000	99.5	10000	61.0

It is evident from these results that the value of μ_∞ for such substances cannot be determined by the method given for strongly dissociated compounds. The dilution at which complete dissociation would take place lies far beyond the possibility of applying the conductivity method directly.

The method of determining the value of μ_∞ for such substances is as follows. While the weak organic acids are only slightly dissociated, salts of these acids are

[1] Ammonia is taken, since work on the substituted ammonias has not generally been carried to very great dilutions.

completely dissociated at moderate dilutions. So also, with respect to the weak bases, which, at ordinary dilutions, are only slightly dissociated; their salts are completely dissociated at dilutions which lie well within the range of the conductivity method. Take an organic acid. Its sodium salt is prepared, and the value of μ_∞ for this salt determined; or taking an organic base, the nitrate of the base is used, and the value of μ_∞ for the nitrate determined.

It remains to see what relation exists between the value of μ_∞ for the sodium salt of an acid, and the acid itself, or between the nitrate of a base and the base.

Kohlrausch[1] has shown that the value of μ_∞ for any compound is the sum of two constants, the one depending upon the cation, the other upon the anion. The value of μ_∞ for sodium acetate is the sum of two constants, the one for the cation, sodium, and the other for the anion, $\overline{CH_3COO}$. If the constant for sodium be subtracted, the remainder is the constant for the anion of acetic acid. If to this constant the constant for hydrogen be added, we have the value of μ_∞ for acetic acid itself. Exactly the same line of reasoning applies to the nitrate of the base.

The constant for $\overline{NO_3}$ is subtracted from μ_∞ for the nitrate, and the remainder is the constant for the cation of the base. To this the constant for hydroxyl is added, and the sum is the value of μ_∞ for the base.

The value of the constant for sodium is 49.2 at 25°, and of hydrogen, 325 at 25°. If we add 275.8 to the value of μ_∞ for the sodium salt of an acid, we have the value of μ_∞ for the acid. The value of the constant for $\overline{NO_3}$, at the same temperature, is 65.1, and for $(\overline{OH})$, 170.

[1] Wied. Ann., 6, 167.

We must, therefore, add 105 to μ_∞ for the nitrate of a base, in order to ascertain μ_∞ for the base itself.

It is thus possible to determine μ_∞ for compounds which are only slightly dissociated at ordinary dilutions. Since μ_v can always be determined for any electrolyte, we are able to measure the dissociation of compounds, which, even in water, are only slightly dissociated.

The application of the conductivity method to measure the exact dissociation in solvents other than water is not always so successful. Water exercises the strongest dissociating action of any known solvent. The ionizing power of many solvents is comparatively so weak that it is impossible to determine the value of μ_∞ for electrolytes, which are strongly dissociated by water, by the direct application of the conductivity method. In such cases, it is possible to determine the dissociation only approximately.

The general applicability of any method to measure electrolytic dissociation is of wide-reaching significance. This will appear, when we consider that many chemical reactions take place between ions, molecules as such not coming into play. The chemical activity of solutions is then a function of the dissociation, and since conductivity is a measure of dissociation, there is a close relation between the conductivity of solutions, and their power to react chemically. Indeed, the former has often been used to measure the latter.

In this connection is to be mentioned, especially, the work of Ostwald[1] on the conductivity of the organic acids, from which he calculated their dissociation constants. Knowing the dissociation constant, the chemical activity of the acid is known. The work of Bredig[2]

[1] Ztschr. phys. Chem., 3, 170, 241, 369.
[2] *Ibid*, 13, 289.

on the conductivity of organic bases is strictly analogous to that just cited.

Ostwald[1] has also shown that it is possible to determine the basicity of acids by determining the conductivity of their sodium salts.

The conductivity method has also been extensively applied to determine what we have already called the constants for the ions, or the relative velocities with which the ions move through their solutions.

A large number of applications of the conductivity method to special problems in dissociation have been made in the last few years, so that it may be said to be one of the most important of all the physical chemical methods.

The Application of the Conductivity Method to the Measurement of Electrolytic Dissociation

When a continuous current is passed through a solution of an electrolyte, the electrodes become quickly covered with gas, or, as we say, become polarized. This increases the resistance to the passage of the current, and interferes with the measurement of the resistance of the solution. Several devices have been proposed for overcoming the effect of polarization,[2] but none have proved as simple as the use of the alternating current. The effect of polarization, tending to retard the flow of the current in one direction, is counterbalanced by the action in the opposite direction, where the polarization current adds itself to the original. This method of measuring the conductivity of solutions we owe to Kohlrausch.

The apparatus employed is sketched diagramatically

[1] Ztschr. phys. Chem., I, 105; 2, 902.
[2] Stroud and Henderson: Phil. Mag., 43, 19.

THE CONDUCTIVITY METHOD

in Fig. 9. J is a small induction coil, with only one or two layers of wire. A larger coil must not be used, since it does not give a sharp tone minimum in the telephone. The coil, tuned to a very high pitch, should be inclosed in a box surrounded by a poor conductor of

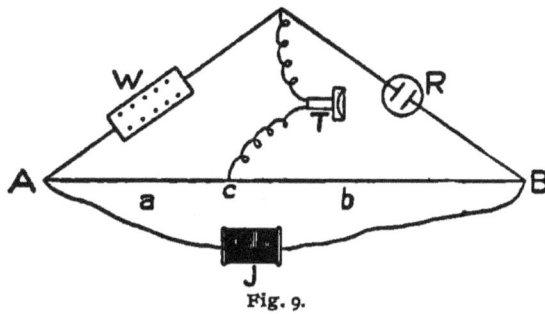

Fig. 9.

sound, and placed at some distance from the bridge where the reading is to be made. The coil is driven by a storage cell of medium size. A platinum wire, or better one of manganese alloy, which has a small temperature coefficient of resistance, is tightly stretched over the meter stick AB, which is carefully divided into millimeters. A rheostat W, whose total resistance amounts to 11,110 ohms, is convenient. The resistance vessel R, containing the solution and electrodes, is shown enlarged in Fig. 10. The electrodes are cut from thick sheet platinum, and into each plate a stout platinum wire, about an inch in length, is welded. Glass tubes are sealed on to the platinum wires and electrode plates, by means of sealing glass, as shown in the drawing. These tubes pass tightly through a rubber cap, which fits over the glass vessel. They are filled to a convenient height with mercury, and electrical connection established by means of copper wires, which dip into the mercury. One arm of the telephone T is thrown into the circuit between the rheostat and the resistance, and the other arm is connected with the bridge wire, by means of a slider.

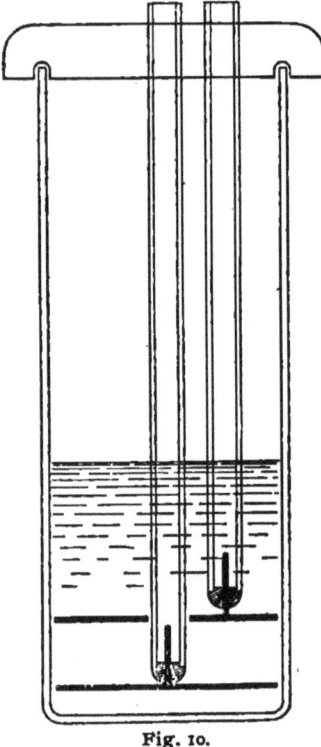

Fig. 10.

This is moved along the wire until that point is found at which the hum of the induction coil ceases to be heard in the telephone. Let this be some point c, and let us represent Ac by a, and Bc by b, the resistance of the solution in the vessel R by r, and the resistance in ohms in the rheostat by w; then, from the principle of the bridge, we have :

$$ra = wb.$$
$$r = \frac{wb}{a}.$$

But the conductivity of a solution c is the reciprocal of the resistance r; therefore,

$$c = \frac{a}{wb}.$$

The conductivity of solutions, determined by this expression, would not, in any sense, be comparable with one another, since there is nothing in the expression which takes into account the concentration of the solution. It is most convenient to refer all concentrations to the molecular normal, containing a gram-molecular weight of the electrolyte in a liter. If we represent by v the number of liters which contains a gram-molecular weight of the dissolved substance, the preceding expression becomes:

$$c = \frac{va}{wb}.$$

Instead of the conductivity c, we write for the molecular conductivity, μ, and to indicate the concentration at which the μ is determined, we write μ_v, in which v has the significance indicated above.

$$\mu_v = \frac{va}{wb}.$$

But even this expression does not take into account the dimensions of the cell used. A cell-constant C must be introduced, and determined for each cell, before it can be employed for conductivity measurements. The complete expression for the molecular conductivity is then

$$\mu_v = C\frac{va}{wb}.$$

Temperature Coefficient of Conductivity

The conductivity of solutions of electrolytes increases rapidly with rise in temperature. The molecular conductivity of a solution of hydrochloric acid, which is 301.7 at 18°, rises to 331 at 25°. This is even more marked in the case of sodium sulphate; a solution having a conductivity of 94.8 at 18° has a conductivity of 171.4 at 50.3°, of 252 at 82°, and of 286.4 at 99.4°.

From this it is evident, that a definite, constant tem-

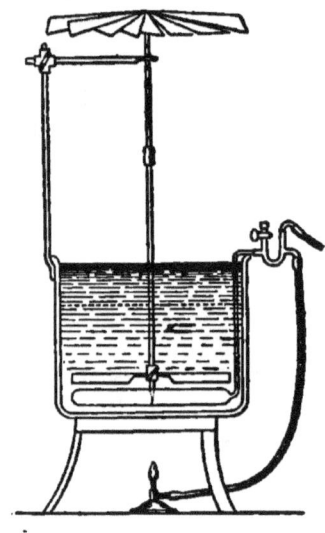

Fig. 11.

perature must be carefully maintained in conductivity work. This is accomplished by placing the vessel containing the solution in a large volume of water, which is maintained at a constant, known temperature. A convenient form of thermostat (Fig. 11) for such work has been devised by Ostwald.[1] A metallic vessel, containing from 15 to 20 liters of water, is stirred by paddles driven

[1] Ztschr. phys. Chem., 2, 565.

by a fan, which is kept in motion by means of a small gas jet beneath. The tube near the bottom of the large vessel is filled with a 10 per cent. solution of calcium chloride. The change in volume of this solution with the temperature, is used to regulate the temperature of the water-bath.

The Ostwald regulator (Fig. 12) can be easily adjusted, so that the temperature of the water-bath will remain constant, to within one-tenth of a degree, for a day. Tube A is connected with the gas supply. The glass tube C, which opens just above the mercury meniscus,

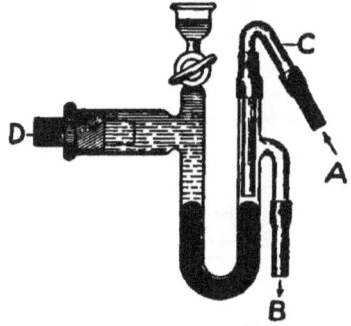

Fig. 12.

contains a fine perforation in the side, so as to supply gas enough to keep the flame alive, when the lower end of the tube is closed by the mercury. Tube B connects with the burner, and D with the large tube containing the calcium chloride solution, resting on the bottom of the water-bath.

When it is desired to adjust the regulator for a definite temperature, the stop-cock is opened, the flame lighted, and a thermometer, divided into tenths of a degree, suspended in the bath. The end of tube C is raised above the mercury surface, and the stirrer is set in motion by means of the small gas jet, placed about a foot below the fans. When the thermometer registers the desired temperature, the stop-cock is closed, and the end of tube C is pushed down until it just touches the mercury surface. The apparatus will then control the temperature automatically.

Calibrating the Wire

A stout platinum wire can be used in constructing the Wheatstone bridge, but, as already stated, it is better to use one of an alloy of manganese (mangandraht). This wire is usually of very nearly uniform resistance, but this can never be taken for granted without testing it. A convenient method for calibrating such a wire has been described by Strouhal and Barus.[1] A piece of German-silver wire about a meter and a half in length, is cut into ten pieces (Fig. 13), which are, as nearly as

Fig. 13.

possible, of the same length. The insulation is removed from the ends of these wires, and they are soldered on to thick copper wires about an inch in length. Nine holes are made in a board, which is about a meter in length, at equal distances apart. These are partly filled with mercury, and receive the ends of the copper wires, which have been previously amalgamated. The board, with the wires in position, is placed along by the side of the bridge wire, and the two end loops attached to the extremities of the bridge. The current from the small inductorium is passed through the bridge, and also

[1] Wied. Annalen, 10, 326.

THE CONDUCTIVITY METHOD

through the series of loops. One of the loops is chosen as the standard of measure, and is suitably marked so as to distinguish it from the others. One end of this standard is attached to one end of the bridge, and the other placed in the first mercury cup. One arm of the telephone is placed in the same mercury cup, and the other attached to the pointer, which moves along the bridge wire. The point of silence on the bridge is ascertained. This is the first reading for point 1. The telephone and all other connections remaining unchanged, the standard measuring wire, which was at position 1, is moved to position 2, and wire 2 is placed in position 1. A reading is again made in the telephone, which is the second reading for position 1. The arm of the telephone, which was in cup 1, is then removed to cup 2, and the point of silence ascertained. This is the first reading for cup 2. The standard wire, which is now in position 2, is moved to position 3, wire 3 is taken back to 2, and all other connections are unchanged. The point of equilibrium is again ascertained at 2, which gives the second reading for this position. The standard wire is thus interchanged in position with each of the loops, and two readings obtained on the bridge for each position except the last, for which only one reading is available.

It must be observed that in all such work in which the telephone is used it is not advisable to try to ascertain directly, the exact point on the wire at which the coil cannot be heard, or at which the tone is a minimum; but to find a point on each side of the true zero, at which the intensity of the tone is the same. These two readings should, at most, be not more than a centimeter apart. The true zero is then just half-way between these points.

The bridge wire is thus divided into ten lengths. The application of the calibration correction is simple. The ten values are added together, and their sum subtracted from 1,000 mm. The difference is divided into 10 parts and each length is corrected by this amount, so that the sum is 1,000 mm. By adding the parts thus, 1, 1 + 2, etc., we obtain the points which correspond to tenths of the wire. The difference between these and 10, 20, etc., gives the correction to be applied.

Carrying Out a Conductivity Measurement

After the wire is calibrated, the next step is to determine the value of the constant (C), for the cell which is to be used. The preparation of the cell is a matter of some care. In the first place, the electrodes must be placed at a convenient distance apart, by shoving the glass tubes through the ebonite cover, and these must then be fastened firmly in the rubber plate, so that no further movement is possible. If a fairly concentrated solution is to be studied, the plates must be as much as 2, or 2.5 cm. apart. If a very dilute solution is to be used, a distance of 0.5 cm. is sufficient. The ordinary white platinum plates, such as are furnished by the manufacturers, cannot be used directly, since they would not give a sharp tone-minimum in the telephone. They must be carefully cleansed by washing in chromic acid, and then in water. A few drops of a solution of platinic chloride are poured into the conductivity cell, (Fig. 10) and the cell filled with pure water until the electrodes are covered. A current from a storage battery is then passed through the solution until the electrodes become more and more deeply blackened. The direction of the current should be frequently altered, so that both electrodes may become coated, and that the deposit may be as uniform as possible. After the plates are completely covered with a layer of the platinum black, the platinic chloride is removed from the cell, a little sodium hydroxide added, and the current passed through this solution. The object of the alkali is to remove any chlorine which may have been retained by the platinum black as it was being deposited. The sodium hydroxide is then removed by hydrochloric acid, and the acid, by repeated washing with pure redistilled water.

In order to determine the value of C, in the expression

$$\mu_v = C \frac{va}{wb},$$

for any cell, it is necessary to use some solution for which the value of μ_v is known. Since potassium chloride can generally be obtained in a high degree of purity, by five or six crystallizations, it is convenient to use in standardizing the cell. A one-fiftieth normal solution of potassium chloride has a molecular conductivity (μ_v) of 129.7 at 25° C. The solution is poured into the cell until the electrodes are covered, and brought to exactly 25° C. in the thermostat. The bubbles of air which usually separate on the electrodes with rise in temperature, having been removed, a resistance is thrown into the circuit by means of the rheostat, which will bring the point of tone-minimum not very distant from the center of the bridge, say between 400 and 600 mm. Thus all the quantities in the above expression, except C, are known, and it can therefore be solved at once for the value of C.

The constant for any given cell being determined, it is a matter of fundamental importance that its value should not be changed. This would be done if the electrodes were moved with respect to one another, or their surfaces in any wise altered. It is therefore necessary that the electrodes should never be placed upon a hard surface, but always upon clean, thick, filter paper, and the plates must never be touched with any hard object.

Knowing the constant for the cell, the measurement of the conductivity of a solution involves exactly the same procedure as that just described. The difference is in the calculation. C is known, and it is desired to find the value of μ_v for a given solution.

The solution is placed in the cell, brought to 25° C., the resistance introduced in the rheostat, and the balance effected on the bridge. All the values in the above expression are now known except μ_v, which is calculated directly.

If the solution used is more concentrated than $\frac{1}{250}$-normal, it is better to use the cell whose electrodes are far apart. If more dilute, the electrodes whose distance from one another is not more than 0.5 cm. should be employed.

Precautions are necessary at every turn. The wire, after calibration, must never be touched with the hand, and the point of contact with the wire must be moved over its surface very carefully. The current must not be allowed to flow through the resistance coils for any considerable length of time, or the temperature, and therefore the resistance of the coils will change. The inductorium should be allowed to run only during the actual measurement of the resistance. Especial care should be taken that every connection is clean and well made, otherwise resistance will be introduced at the junctions.

Correction for the Conductivity of Water

Since water is the most general solvent known, and solutions in this solvent have the greatest conductivity, one is called upon, most frequently to measure the conductivity of aqueous solutions. In all such cases the quantity actually measured is the sum of the conductivities of the water and of the dissolved electrolyte. The conductivity of the water alone, must, in every case, be determined, in order that the conducting power of the electrolyte may be ascertained. It would, at first sight, appear to be possible to use water of only a fair degree of purity, to determine its conductivity, and then to subtract this from the conductivity of the solution. Whether this could be done, would depend upon the nature of the impurities.

They might easily be of such a character as to react chemically with the dissolved electrolyte, and thus seriously affect the nature of the solution. Thus, ammonia, which would neutralize any acid, forming a salt, would materially change the nature of the ions present, and therefore the conductivity. Carbon dioxide would, in like manner, affect the conductivity of any strong base. It is therefore necessary, in all work involving the use of the conductivity method, to prepare water in as pure condition as is practicable, and then to introduce a correction for its conductivity, when this is larger than the necessary experimental error.

Kohlrausch[1] has prepared the purest water thus far obtained, by distilling the purest water obtainable by other methods, in a vacuum. He determined its conductivity without exposure to the air, and found it to be 0.04×10^{-6}. To prepare water of this degree of purity is not practicable, and indeed is not necessary for conductivity work.

Nernst[2] has suggested fractional crystallization as a

[1] Ztschr. phys. Chem., 14, 317.
[2] Ibid, 8, 120.

means of purifying water for conductivity purposes, but equally efficient and far more rapid methods have been subsequently devised.

Hulett[1] has obtained water of a high degree of purity, by distilling it first from potassium bichromate and sulphuric acid, and then redistilling from a solution of barium hydroxide. The water purified in this way had a conductivity of from 0.7 to 0.8 $\times$ 10^{-6}.

More recently, Jones and Mackay[2] have used an apparatus in which the water is distilled first from acid potassium permanganate or acid potassium bichromate, which decomposes any organic matter present and retains the ammonia, and second from alkaline potassium permanganate, which retains any carbon dioxide. The apparatus is shown in Fig. 14. Ordinary

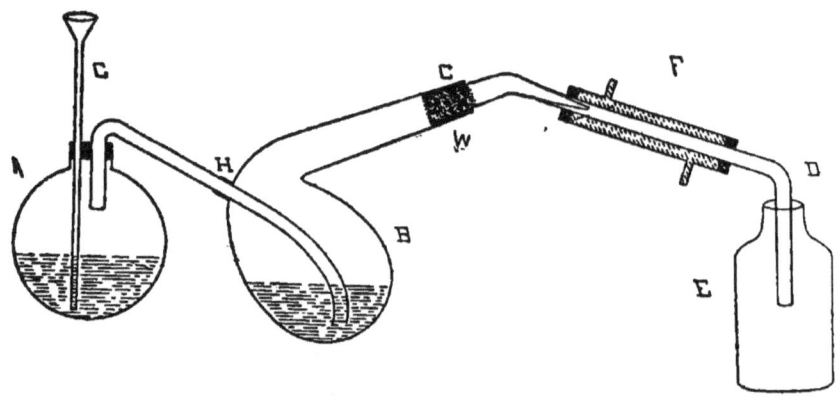

Fig. 14.

distilled water is introduced into the vessel A, together with a little sulphuric acid and potassium permanganate, or potassium bichromate, through the

[1] Ztschr. phys. Chem., 21, 297.
[2] Am. Chem. J., 19, 91; Ztschr. phys. Chem., 22, 237.

THE CONDUCTIVITY METHOD 63

funnel tube G. It is heated to boiling, the vapor passing into B, which contains distilled water, potassium permanganate, and a little potassium or sodium hydroxide. A small flame is sufficient to keep the liquid in B at the boiling temperature. The vapor passes from B along the long neck of the retort, over the glass wool, which is meant to arrest any trace of permanganate carried along by the steam, into the tin condenser, and is received in the flask E. Certain precautions must be taken in fitting up and using the apparatus. The glass wool W, introduced into the adapter arm C, must fill only the lower part of the arm, otherwise there is danger that a trace of alkali dissolved from it, will be swept over into the condenser. The glass wool should be washed well with hydrochloric acid. Whenever the apparatus is cleaned and refilled, which should be done about once a week when in constant use, the distillate collected at first, must be discarded, since it always contains a trace of alkali, probably of ammonia, formed by the action of the alkaline permanganate on the organic impurities in the ordinary distilled water introduced into B. When this is once removed, there is no further escape of ammonia possible, since the organic impurities in the water are destroyed by the permanganic acid in A, and the ammonia combines with the sulphuric acid present in that vessel. The carbon dioxide liberated in A, is absorbed by the alkali in B. The process is thus perfectly continuous for at least a week, or it can be interrupted at any time, by removing the burner. Four or five liters of water can be obtained daily with the use of this apparatus.

The water purified by this method gave a conductivity at $25°$, varying from 1.5 to 2.0×10^{-6} in mercury units.

The correction which must be applied to the values of

μ_v, for the conductivity of the water employed in preparing the solutions, is calculated by multiplying the specific conductivity of the water by the molecular volume of the solution in cubic centimeters. This quantity, for water properly purified, is negligible for all ordinary concentrations, and attains an appreciable value only in dilute solutions. In case the substance under investigation reacts chemically with the impurities in the water, such a correction would be so uncertain that it is better not to attempt to apply it.

Substances to be Used

In practice, it is well to use some of the same substances whose freezing-point lowerings have been measured, that the dissociation as determined by conductivity, may be compared with that calculated from the depression of the freezing-point. Prepare say a tenth-normal solution of the substance chosen, determine the value of μ_v for this dilution, increase the dilution to $\frac{1}{100}$-, $\frac{1}{500}$-. $\frac{1}{1000}$-, $\frac{1}{5000}$-, and $\frac{1}{10000}$-normal, determining in each case the value of μ_v. At about $\frac{1}{1000}$, μ_v will become constant, and will show no further increase with increase in dilution. This is the value of μ_∞. To find the percentage of dissociation, α, at any dilution, divide the value of μ_v at that dilution, by the value of μ_∞ for the substance.

$$\alpha = \frac{\mu_v}{\mu_\infty}.$$

www.ingramcontent.com/pod-product-compliance
Lightning Source LLC
Chambersburg PA
CBHW020252090426
42735CB00010B/1896